Field Identification *of* Coastal Juvenile Salmonids

by

W.R. Pollard
G.F. Hartman
C. Groot
Phil Edgell

Illustrated by C. Groot
Photography by Phil Edgell

This field guide has been prepared cooperatively by the Department of Fisheries and Oceans and Weyerhaeuser Ltd.

Net proceeds from the sale of this publication will be donated to the **Pacific Salmon Foundation**, a non-profit charitable organization which supports community-based volunteer groups dedicated to the enhancement and restoration of BC's salmon resources.

HARBOUR PUBLISHING

Published by Harbour Publishing for the Federal Department of Fisheries and Oceans and Weyerhaeuser Ltd.

Harbour Publishing
P.O. Box 219
Madeira Park, BC Canada V0N 2H0

Cover, page design, maps and composition by
Martin Nichols, Lionheart Graphics.
Cover photo by Rick Blacklaws.
Printed and bound in Canada.

CANADIAN CATALOGUING IN PUBLICATION DATA

Main entry under title:

Field identification of coastal juvenile salmonids

ISBN 1-55017-167-4

1. Salmonidae—Identification. I. Pollard, W.
QL638.S2F53 1997 597.5'5 C97-910344-4

Acknowledgements

Many people contributed helpful ideas and information or assisted with collection of live fish used in the photographs and drawings in this guide, including: Rick Axford, Ian Baker, David Barnes, Ray Billings, E. Groot, W. Hartman, L.B. Holtby, A. Lamb, J.D. McPhail, Tony Massy, Skip Rimmer, Al Ross, Lanny Ross, Cole Shirvell, and Herman Watts. Stan Coleman, Weyerhaeuser in Nanaimo, and Don Lawseth, Department of Fisheries and Oceans in Vancouver, contributed time of employees to complete this project. We are especially indebted to Barb Knight, Weyerhaeuser in Nanaimo, who took our ideas and moulded them into the format used in the guide. Blayney Scott of Scott Plastics, a frequent contributor to salmon enhancement projects, provided the frames and bags.

Table of Contents

INTRODUCTION

Correct identification of young salmonids will improve the accuracy of resource management information and lead to a fuller knowledge of the distribution and status of fish stocks. This guidebook will help you to identify young salmonids in the field. Fish identification requires practice, but learning to identify young salmonids can be an enjoyable and worthwhile endeavour.

SCOPE

Information is provided for 10 species of juvenile salmonids found in coastal BC watersheds. Regional differences occur in the appearance of fish, and their physical features may change as they grow. The range of characteristics is vast; this guide is limited to the freshwater rearing stages.

We have chosen to exclude brook trout *(Salvelinus fontinalis)* from this guide to avoid confusion with the widely distributed Dolly Varden *(Salvelinus malma)*, and due to the fact that brook trout are known to be present in so few areas on the BC coast. Other field guides should be consulted when sampling in a location known to have brook trout.

Interior regions of BC have salmonid species or races that are not found in coastal waters. You should exercise caution when using this guide for interior waters or species.

TAXONOMY OF BC SALMONIDS

Salmon *(Oncorhynchus)*, trout *(Salmo)* and char *(Salvelinus)* are the three genera of the family Salmonidae found in coastal BC fresh waters. The scientific and common names of BC coastal salmonids are:

Pink salmon *(Oncorhynchus gorbuscha)*
Chum salmon *(O. keta)*
Sockeye salmon *(O. nerka)*
Chinook salmon *(O. tshawytscha)*
Coho salmon *(O. kisutch)*
Cutthroat trout *(O. clarki)*
Steelhead/rainbow trout *(O. mykiss)*

Brown trout *(Salmo trutta)*
Atlantic salmon *(S. salar)*

Dolly Varden *(Salvelinus malma)*
Bull trout *(S. confluentus)*
Brook trout *(S. fontinalis)*

Common names are often not consistent with the scientific names. The Atlantic salmon *(Salmo salar)* is a trout. Cutthroat and rainbow trout *(Oncorhynchus clarki and O. mykiss)* have recently been placed in the salmon genus. Bull trout *(Salvelinus confluentus)* are char. We have adopted the common names used by McPhail and Carveth (1994).

Coastal BC fresh waters have relatively few native species and the fish most often encountered are salmonids. However, considerable biological diversity occurs within each species. These physical and behavioural differences in salmonids promote survival in the large variety of BC's coastal environments. For example, a resident rainbow trout in a small headwater reach of a stream may spend its entire life near where it emerged from the gravel and never grow larger than 15 cm. The anadromous form of rainbow trout (steelhead) will rear in fresh water for up to several years and then migrate up to several thousand kilometres at sea, where it can grow to more than 1 m and 10 kg before returning as an adult. Both forms are rainbow trout but their life histories are very different.

Identification Procedures

Users of this guide will find several pathways to identify juvenile salmonids. They include identification charts, colour and black and white illustrations, colour photos of live fish, and distribution information.

The Identification Charts can be used to determine whether fish are salmon, trout or char and the individual species. To confirm identification, the drawings and information in the Detailed Species Information pages provide summaries of physical features and distribution. The generalized distribution information is useful in determining when and where in watersheds the user can expect to find the various species. The photos show examples of how the features are expressed on live fish.

Pairs of species with similar physical characteristics (rainbow/cutthroat, coho/chinook, chum/sockeye) are positioned next to each other to facilitate correct identification.

Few fish can survive being held out of water while the guidebook user goes through the identification steps. Please use the fish viewing bag provided with the guidebook to keep fish alive while identifying them. Do not overload the bag and change the water regularly to keep your specimens healthy. Careful handling will allow you to identify most fish without harming them. We recommend that you use a 10x magnifying glass to see some of the features on small fish.

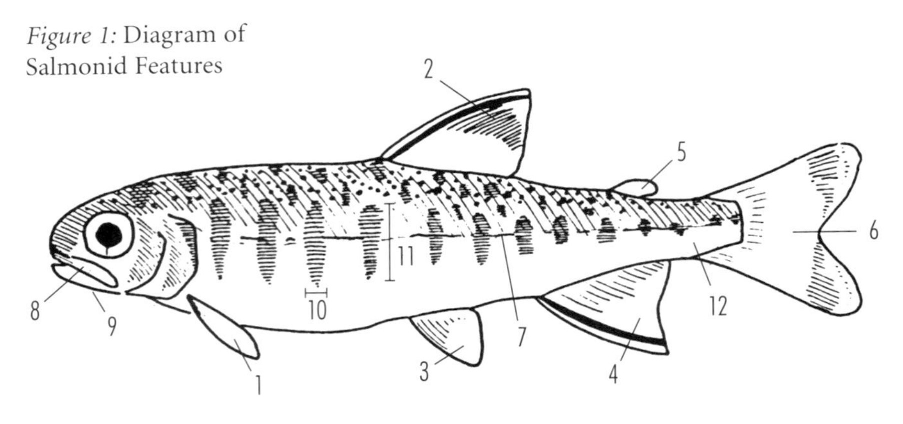

Figure 1: Diagram of Salmonid Features

Key:

1. Pectoral fin
2. Dorsal fin
3. Pelvic fin
4. Anal fin
5. Adipose fin
6. Caudal fin
7. Lateral line
8. Maxillary
9. Branchiostegals
10. Parr mark width
11. Parr mark height
12. Caudal peduncle

Salmonid identification characteristics vary over geographic areas. The illustrations used in this guide generalize and emphasize those characteristics. The photos of live fish from Vancouver Island waters show examples of what you may see in the field. Features and colours vary greatly among fish. Stress may affect the intensity of colours. These things make it necessary to look carefully at live fish to see features that are clear on drawings. Some of the key features that are usually easy to see on live fish are listed with the photos.

What if I use the guide and still can't identify the fish?

Some juvenile fish are difficult to identify. Even experienced biologists occasionally have trouble with the field identification of some small salmonids. For example, determining the difference between rainbow and cutthroat trout that are smaller than about 80 mm is very difficult. Fortunately, continued sampling of species that rear in fresh water for several years will usually yield larger specimens that are easier to identify. If you can't identify a specimen, then report the fish as "unknown" or identify the fish as far as you can confidently take it (e.g. trout, char, etc.).

Sampling Rules & Regulations

Capturing juvenile salmonids for identification purposes requires sampling permits from the Department of Fisheries and Oceans and the BC Ministry of Environment, Lands and Parks. Some characteristics involving detailed counts or other procedures require that fish be preserved. Check the conditions of your permit to determine whether you can retain fish and do so only if it is absolutely necessary to confirm your identification. Examining fish for features highlighted in red in the Identification Charts or on the Detailed Species Information pages can seriously harm or kill the specimen. Check your permit before attempting to use these diagnostic features.

Figure 2

Identification Chart

Salmon and trout/char

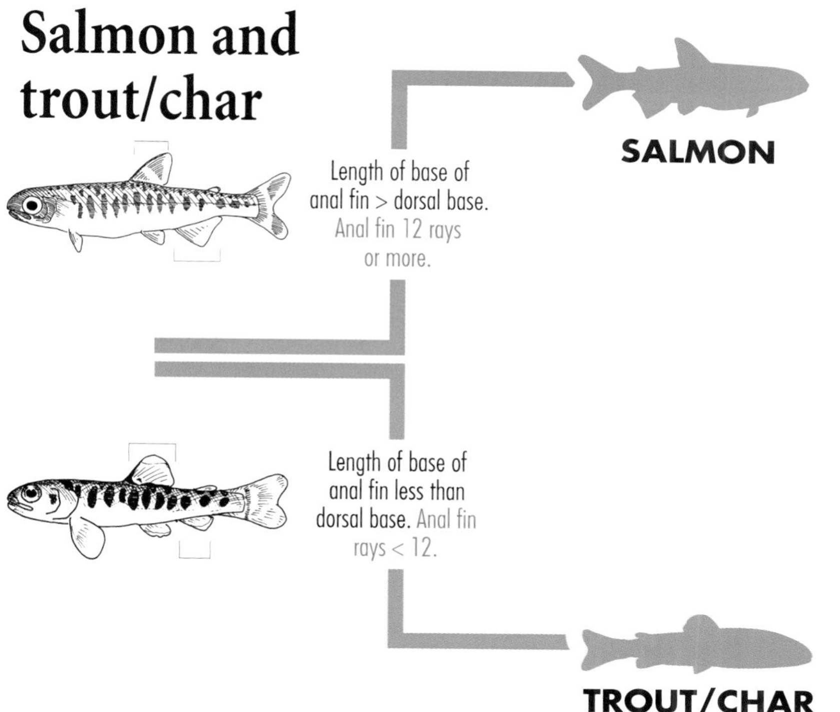

● Examining fish for features marked in red can harm or kill the specimen, and may require 10x or binocular microscope.

Figure 3

IDENTIFICATION CHART

Five species of juvenile Pacific salmon

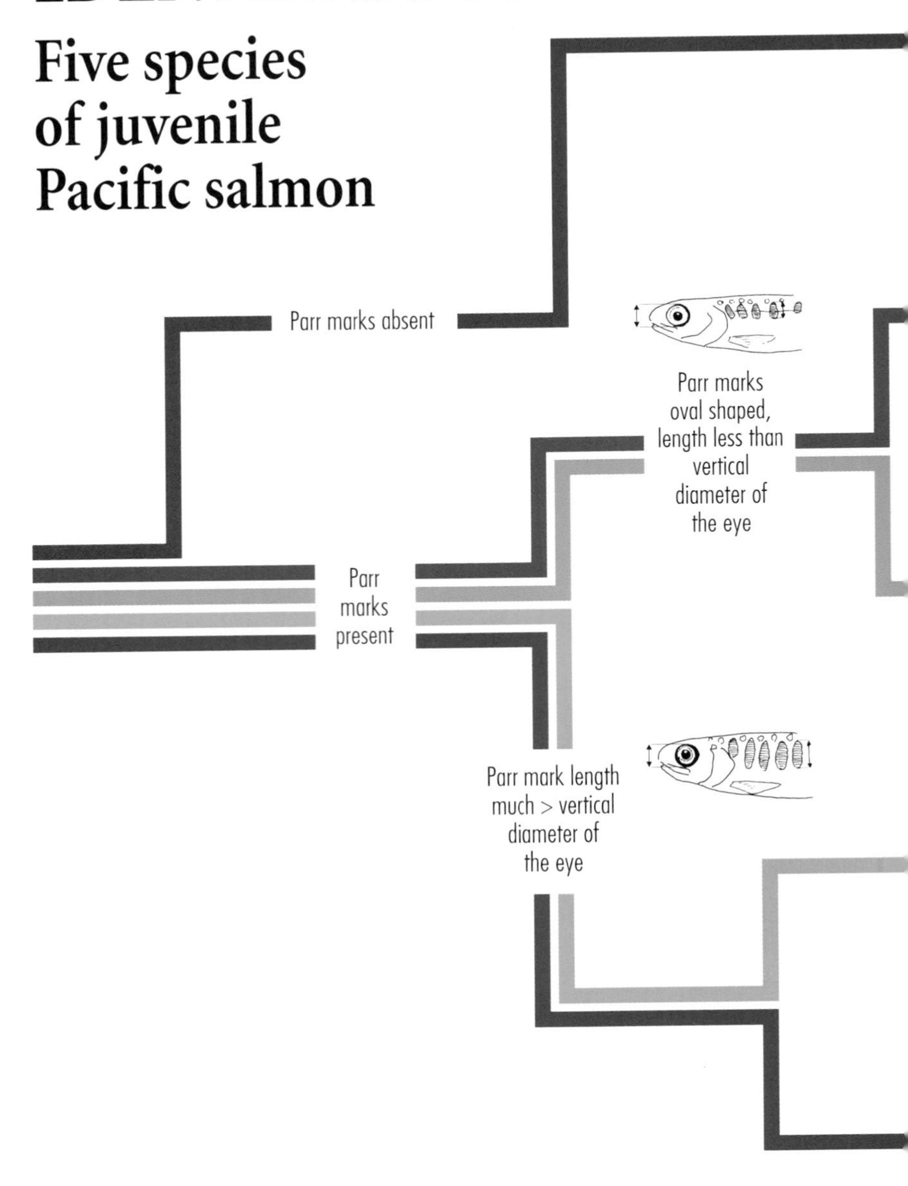

● Examining fish for features marked in red can harm or kill the specimen, and may require 10x or binocular microscope.

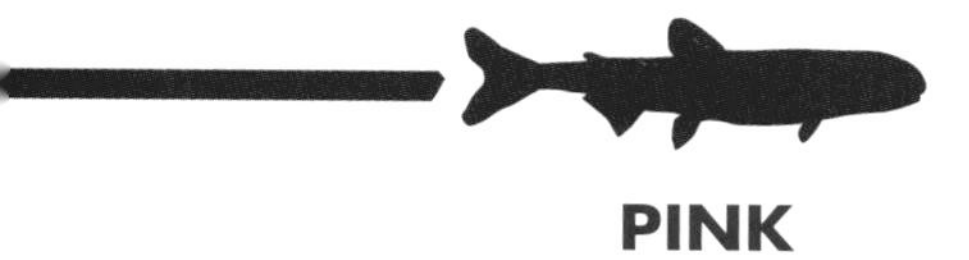

PINK

1 Bright silvery colour.
2 Going to sea usually at < 40 mm.

Dorsal surfaces
mottled green,
ridescent green
elow lateral line,
arr marks regular
in shape and
osition on sides

CHUM

1 Size in fresh water usually < 50 mm.
2 Fish < 50 mm have mottled green back and silvery sides.
3 Parr marks faint or absent below lateral line.
4 Light areas between parr marks on lateral line, < 2x parr mark width on average.
5 19–26 gill rakers on first arch. Gill rakers stubby, about half the length of gill filaments.

No iridescent
green below
ateral line, parr
marks slightly
regular in shape
and position

SOCKEYE

1 Size in fresh water to 100 mm.
2 Some parr marks, about half above and half below lateral line.
3 Light area between parr marks on lateral line, about 2x average width of parr marks.
4 30–39 gill rakers on first arch, gill rakers long and slender, = to or > length of gill filaments.

oose fin uniformly
igmented, anal
ickle-shaped, anal
and dorsal fins
th white leading
dge followed by
a black stripe

COHO

1 Leading edge of anal fin longer than base.
2 Orange hue, particularly on adipose, anal and caudal fins.
3 13–14 branchiostegals.
4 Usually 45–80 pyloric caeca.

dipose fin with
unpigmented
indow, anal fin
ot sickle-shaped,
rsal fin with dark
ading edge and
white tip

CHINOOK

1 Leading edge of anal fin < the length of base.
2 16–18 branchiostegals.
3 Usually 135–185 pyloric caeca.

Figure 4

Identification Chart

Juvenile trout and Dolly Varden

Pectoral fin as long as depressed dorsal fin, and reaches to a vertical line through the anterior insertion of the dorsal fin

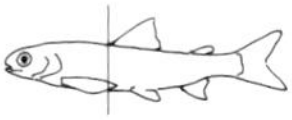

Parr marks large and irregular in shape and position, dorsal leading margin faintly edged in black, vomerine teeth on the head of vomer only

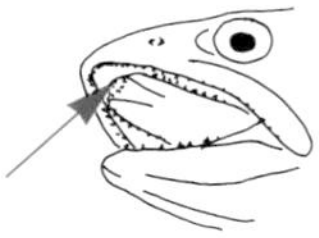

Pectoral fin not as long as depressed dorsal fin, and does not reach to a vertical line through anterior insertion of dorsal fin

Parr marks regular in shape and position, distinct dark spots once fish reach 8 cm, teeth on head and shaft of of vomer

Adipose fin not orange, distinct dark pigment on lower part of first dorsal fin ray in fish 40 mm. or less

Note: You may encounter juvenile mountain whitefish in some mainland rivers, especially those that penetrate the coastal mountains to interior regions. Juvenile mountain whitefish have an adipose fin and parr marks; however, their bodies are more slender and pencil-shaped than trout or char. The mouth of a juvenile whitefish is positioned lower down on the jaw and is smaller than that of trout or char. Juvenile mountain whitefish have fewer and thus larger scales than trout or char. Whitefish have 70–90 oblique rows of scales across the lateral line, compared to 100 or more rows on trout or char.

⋆ Need for 10x hand lens or binocular microscope.

■ Melanophores are small black pigment cells about the si fine pepper. Spots on fish are an accumulation of melanophores. Small black melanophores often show or recently emerged fry but these are not what we refer to spots.

ATLANTIC

1 Red dots on lateral line on larger fish.
2 Adipose fin not orange.

DOLLY VARDEN or BULL TROUT

1 No black spots on back or sides. ■
2 Width of parr marks on lateral line greater than light areas.
3 Small triangle-shaped pigment spot at base of caudal fin.*

dipose fin orange, orange spots on or near lateral line in larger juveniles

BROWN TROUT

1 Adipose fin orange.
2 Small black spots above and below lateral line.
3 9–12 parr marks, greater than diameter of eye.
4 Orange spots, if present, are close to lateral line.

axillary reaches past he posterior margin of the eye in fish over 8 cm. Red or range hyoid colour sually present once sh reach 8–10 cm

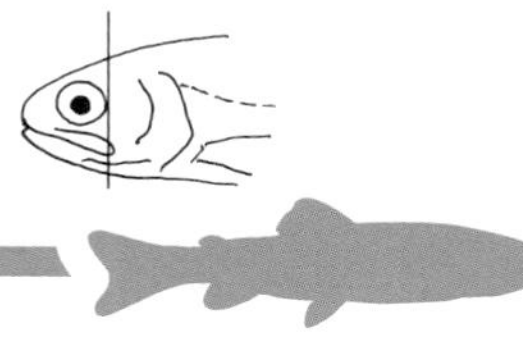

CUTTHROAT

1 Caudal fin melanophores tend to form in lines along fin rays in fish < 50 mm.* ■
2 Mid-dorsal parr-like marks usually absent.
3 White tip on dorsal covers 3 or fewer ray interspaces.
4 Maxillary reaches past posterior margin of eye (does not separate in trout less than 8 cm).
5 Hyoid teeth present.

Fish over 8 cm axillary reaches to ack of eye or less, no hyoid red or orange on any uvenile fish, even se over 10 cm long

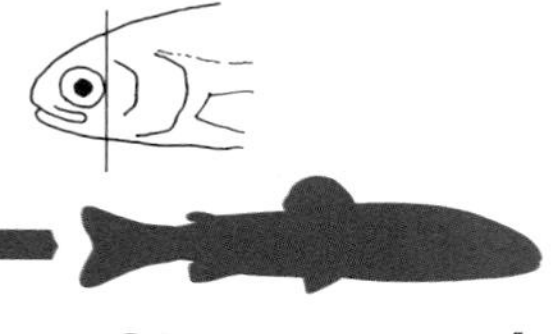

STEELHEAD/ RAINBOW

1 Caudal fin melanophores are evenly distributed in fish < 50 mm.* ■
2 Fish > 50 mm. median dorsal parr marks usually present.
3 White tip of dorsal covers 3–5 ray interspaces.
4 Hyoid teeth absent.

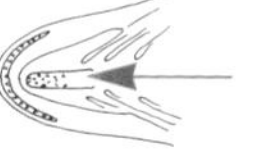

Examining fish for features marked in red can harm or kill the specimen, and may require 10x or binocular microscope.

Dolly Varden & Bull trout

Salvelinus malma and *S. confluentus*

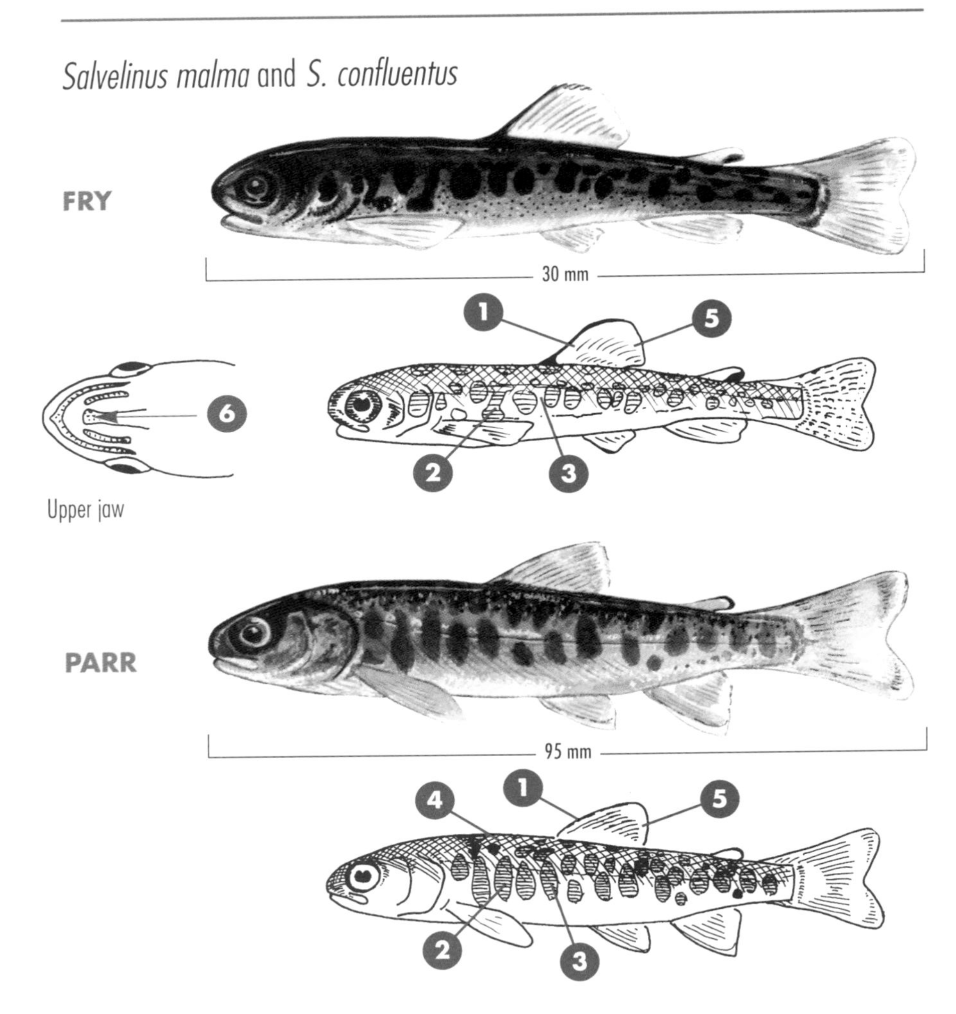

Colour & Anatomy

1 Dorsal fin—leading edge some pigment, but first ray is not heavy black *(see trout)*.
2 Parr marks are irregular shape and location.
3 Dark on midline is much greater than light areas.
4 Back and sides have no black spots (recently emerged fry may have melanophores).
5 Fins have no black spots.
6 Only head of vomer has teeth.

Distribution

- Dolly Varden are widespread in coastal waters. Bull trout are more commonly found in interior BC waters, but are occasionally found in some coastal mainland waters.
- Populations live in large and small permanent streams and lakes.
- Dolly Varden and bull trout may use small tributaries and upstream reaches and may co-exist with cut-throat or rainbow trout.
- Dolly Varden and bull trout may exist in isolated populations above stream barriers.
- Anadromous Dolly Varden are uncommon in some areas of Vancouver Island. Resident forms are widely distributed on the BC coast.

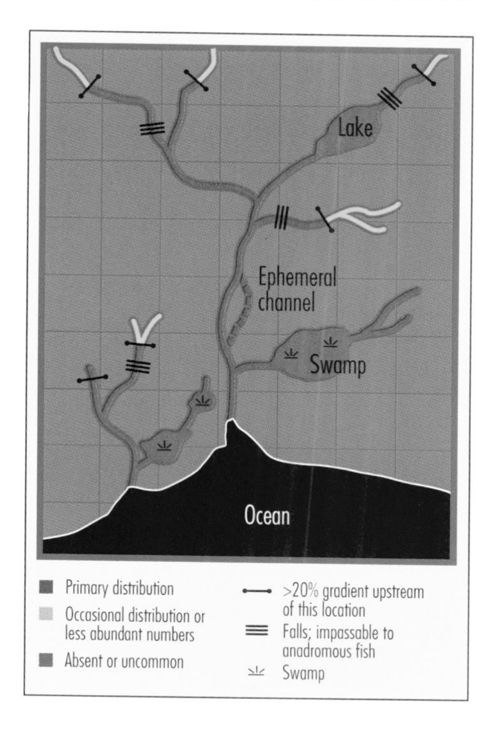

Behaviour

- Populations may be resident or anadromous.
- Dolly Varden and Bull trout may hybridize.
- Dolly Varden and bull trout spawn during fall.

Freshwater residence time

- Anadromous forms will rear for up to several years before migrating to sea.

Other

- Small Dolly Varden are difficult to distinguish from bull trout.
- Specimens larger than 10–12 cm will have light spots on a dark background. May have dark parr marks, but will not have black spots on their bodies or fins.

CUTTHROAT TROUT

Oncorhynchus clarki

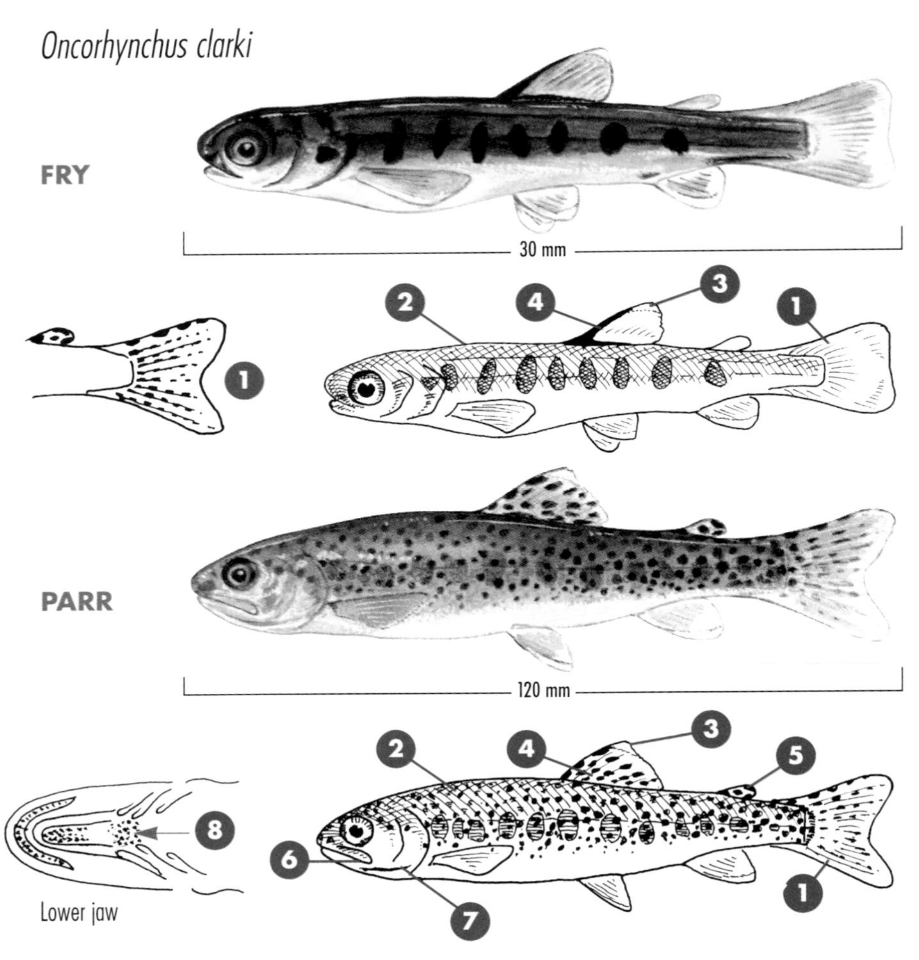

COLOUR & ANATOMY

1 Melanophores are in spots or streaks along rays in caudal fin of fry <50 mm. (Note: magnifying glass may be required to observe this trait.)
2 Median–dorsal parr-like marks are usually absent.
3 White tip on dorsal covers 1 to 3 interspaces between dorsal fin rays.
4 First ray is black on fry.
5 Adipose may have 1–2 breaks in pigment on rim and often spotted on parr.
6 Maxillary extends past rear margin of the eye on fish >80 mm.
7 Underside of jaw (on parr) has red or yellow slash.
8 Hyoid teeth are present at the base of the tongue behind first gill arch—see inside lower jaw.

Distribution

- Populations live in large and small permanent lakes and streams.
- Resident fish may be found in lakes and streams.
- Cutthroat are usually farther up in the system than steelhead/rainbow.
- Cutthroat may coexist with Dolly Varden in small stream tributaries.
- Cutthroat may occur above barriers in streams.
- Cutthroat use off-channel habitats such as intermittent tributaries and sloughs.
- Cutthroat may also occur in small tributaries upstream from sloughs—steelhead do not.

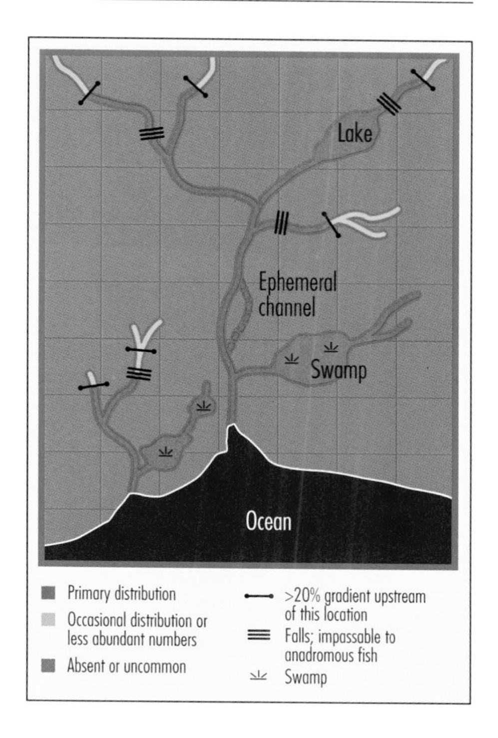

Behaviour

- Both anadromous and resident forms occur.
- Cutthroat hide under logs and debris piles, or under streambank at temperatures <6°C.
- Cutthroat may occasionally spawn with rainbow and produce hybrids.
- Cutthroat spawn from late winter through spring.

Freshwater residence time

- Anadromous fish are resident up to 3 years.
- Resident fish spend their entire life in fresh water.

Other

- Small cutthroat are very difficult to distinguish from steelhead/rainbow, and features may be similar on fish <10–12 cm long.
- Continued sampling will usually yield fish >10 cm that are more easily identified than younger fish.

STEELHEAD/ RAINBOW TROUT

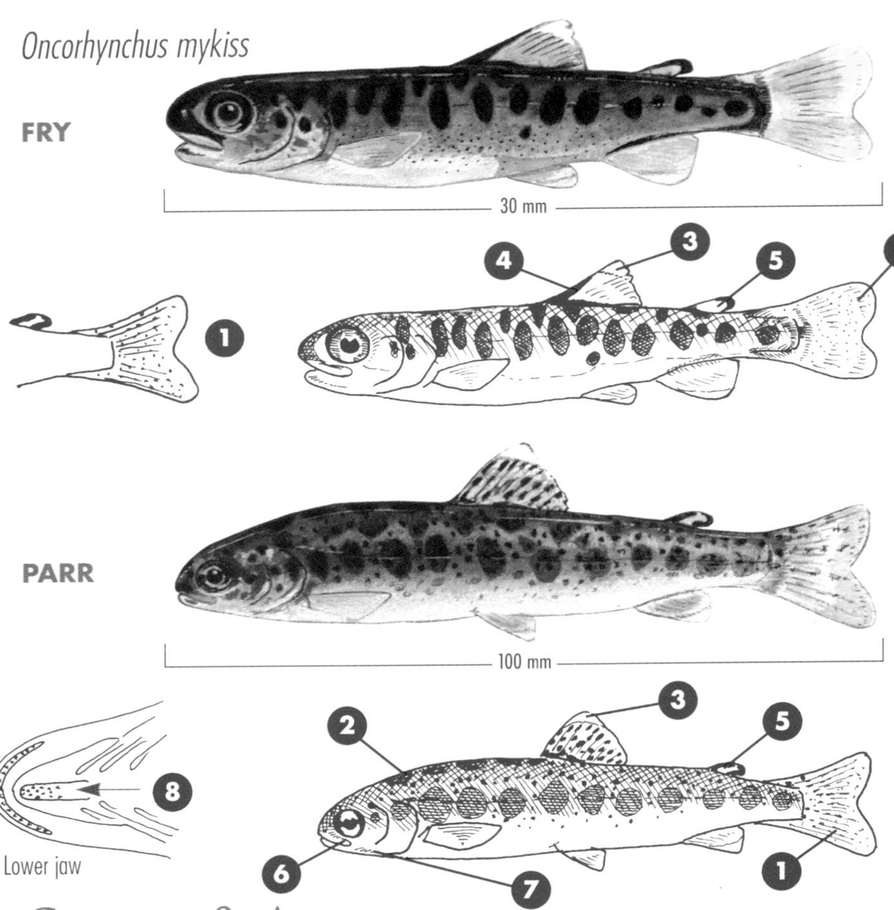

COLOUR & ANATOMY

1 Melanophores are evenly speckled on caudal fin of fry. (Note: magnifying glass usually required to observe this trait.)
2 Median–dorsal area has parr-like marks, about 5.
3 White tip on dorsal covers 3 to 5 interspaces between dorsal fin rays.
4 First ray is black on fry.
5 Adipose usually has continuous rim of pigment or one break.
6 Maxillary does not extend past back margin of eye of parr.
7 Jaw has no red or yellow slash.
8 There are no hyoid teeth.

Distribution

- Steelhead/rainbow are found in main channel, permanent tributaries and lakes.
- Steelhead/rainbow can use small permanent streams and stable side channels.
- Small resident forms may be found isolated above barriers and in lakes.
- Rainbow do not normally coexist with cutthroat in headwaters. Steelhead/Rainbow and cutthroat are found together in anadromous waters and occasionally coexist in large rivers and lakes.
- Finding both rainbow and cutthroat in streams usually signifies anadromous use.
- Resident and anadromous forms can be distinguished by opening fish and checking for maturity. Mature males/females in the 115–120 mm. range signify resident fish. If fish are immature at this size, assume they are juvenile steelhead.
- Large lakes can have late spawning populations where fry do not emerge until late summer.

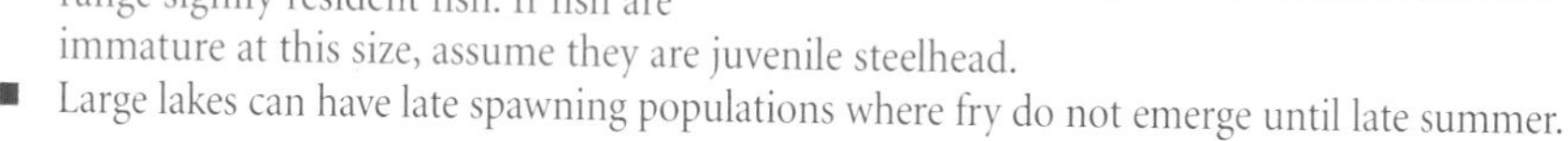

Lake
Ephemeral channel
Swamp
Ocean

Primary distribution
Occasional distribution or less abundant numbers
Absent or uncommon
>20% gradient upstream of this location
Falls; impassable to anadromous fish
Swamp

Behaviour

- Steelhead/rainbow hide in streambed or debris piles at temperatures <about 6°C. They may move out at night.
- Steelhead/rainbow may occasionally spawn with cutthroat and form hybrids.
- Steelhead/rainbow usually spawn from late winter through spring.

Freshwater residence time

- Anadromous fish rear for up to 3 years before going to sea.
- Resident fish spend their entire life in fresh water.

Steelhead/Rainbow vs Cutthroat

- Steelhead/rainbow are less likely to be found in ephemeral off-channel habitat or tributaries above sloughs or swamps.
- Steelhead/rainbow maxillary does not extend past back of eye, hyoid teeth and red slash are absent.
- Cutthroat maxillary extends past the eye, hyoid teeth are present, red slash is present on bottom of the jaw.

BROWN TROUT *(introduced)*

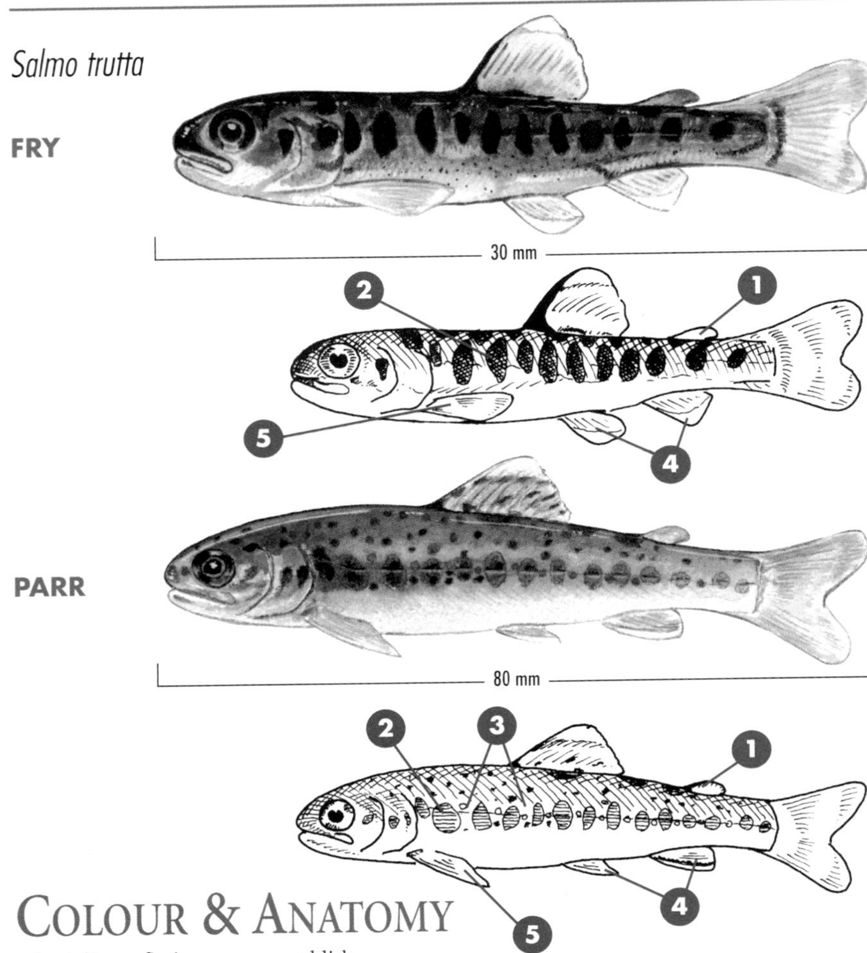

COLOUR & ANATOMY

1 Adipose fin is orange or reddish.
2 Species has 9–12 parr marks.
3 Red or orange spots show along or near lateral line of parr.
4 Anal and pelvic fins may have white leading edge.
5 Pectoral fins are wide, but do not extend back to vertical line through front of dorsal.

DISTRIBUTION

- Brown trout are introduced in Cowichan, Little Qualicum and Adam (Vancouver Island) rivers, including some lakes in those watersheds.

OTHER

- Adults have light halos around spots.
- Brown trout spawn during fall.

ATLANTIC SALMON *(introduced)*

Salmo salar

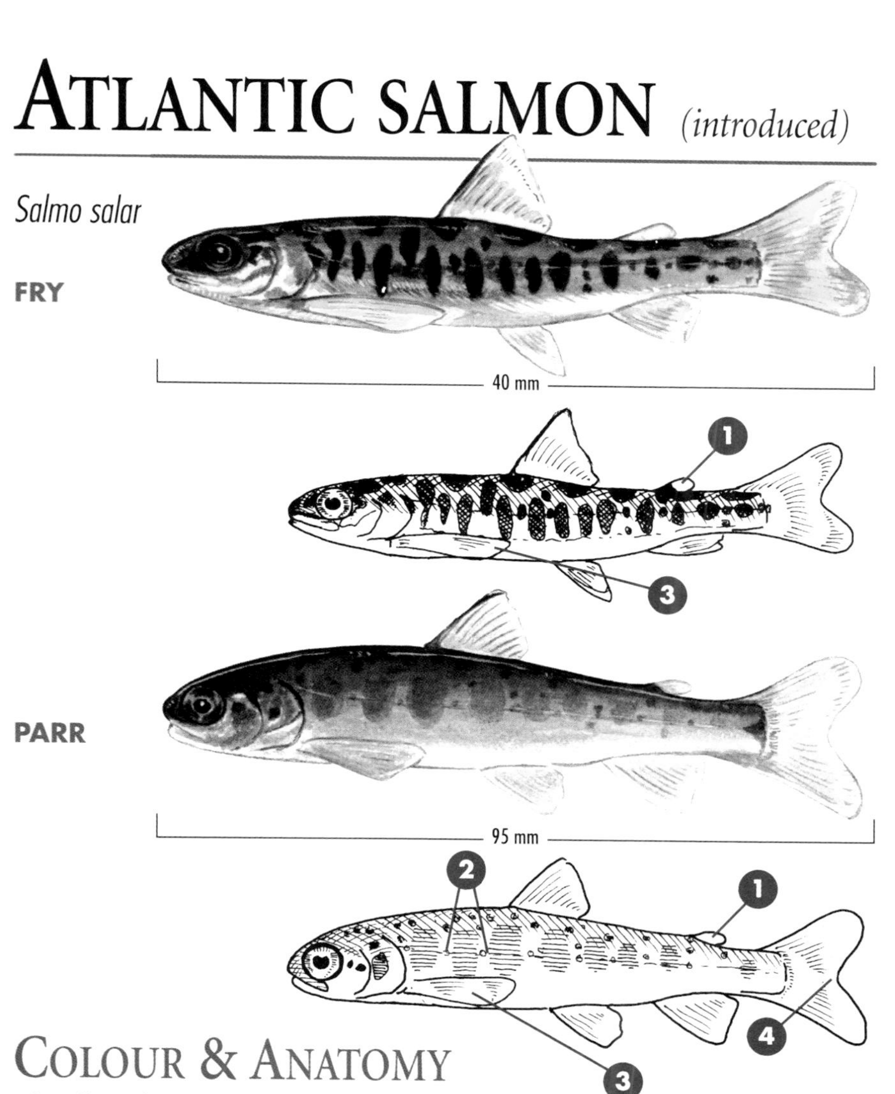

COLOUR & ANATOMY

1. Adipose fin is clear.
2. Red spots are regular between parr marks on lateral line of parr.
3. Pectoral fin is long, reaches to or past a vertical line through front of dorsal fin.
4. Caudal fin of parr is more deeply forked than in brown trout parr.

Note: Save any fish which match the above description and submit your specimen to the nearest Ministry of Environment, Lands and Parks regional office. Atlantic salmon are similar to brown trout. Check your ID if fish is captured in waters known to contain brown trout.

DISTRIBUTION

- Distribution not known. Fish are escapees from net pens at fish farms. They may occur near escape sites, particularly middle to north Vancouver Island. It is not known whether they are reproducing naturally.
- Adults are taken by fishermen in the sea and some have been caught in rivers.

PINK SALMON

Oncorhynchus gorbuscha

FRY

30 mm

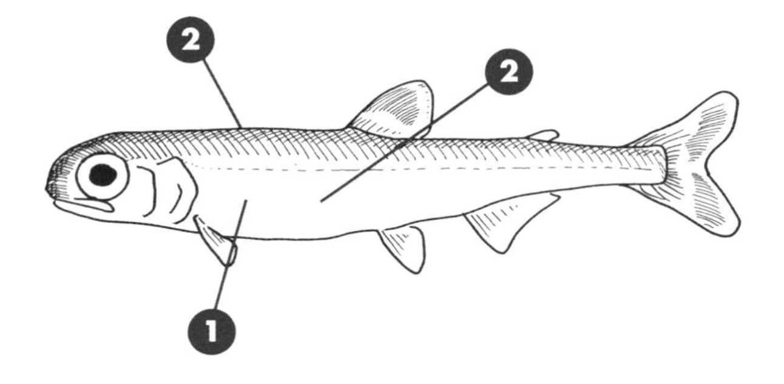

COLOUR & ANATOMY

1 Parr marks are absent.
2 Dorsal surface is green, ventral is silver.

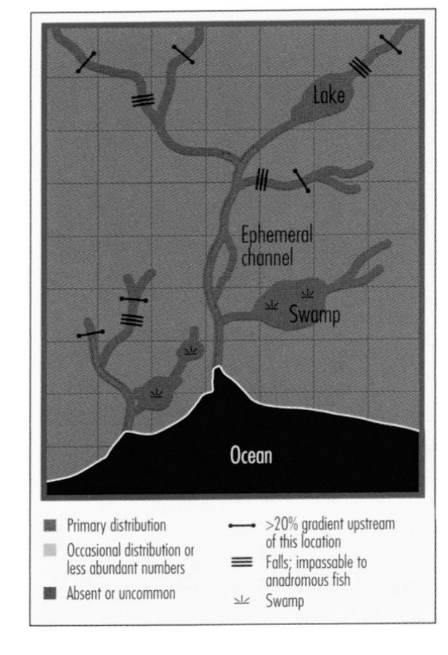

DISTRIBUTION

- Pink salmon are found in lower reaches of small coastal streams.
- Some runs ascend large rivers for several hundred kilometres.

FRESHWATER RESIDENCE TIME

- Fry hatch, emerge and go directly to sea.

OTHER

- Fry are 3–3.5 cm. long—maximium size in freshwater 4.5–5 cm.
- Pink salmon may spawn in the intertidal zone.
- All adult pink salmon are 2 years old. Rivers frequently have adult runs only every other year. A few watersheds have natural runs every year, but the even-numbered year or odd-numbered year run will typically be stronger.

CHUM SALMON

Oncorhynchus keta

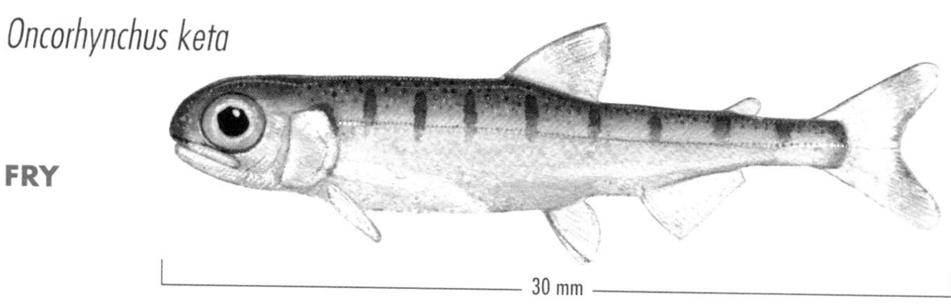

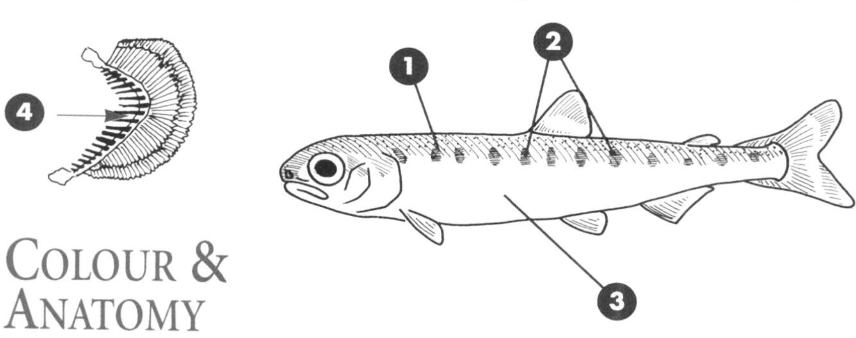

COLOUR & ANATOMY

1. Parr marks smaller than vertical diameter of eye, and faint or absent below lateral line.
2. Parr mark height is more regular than on sockeye.
3. Area below lateral line has pale greenish iridescence.
4. Gill rakers are short and stubby, about half the length of the gill filament, 19 to 26 on first gill arch.

DISTRIBUTION

- Chum usually occur in lower reaches of most coastal streams.
- Chum migrate up to 160 km up Fraser River and lower reaches of large tributaries. Fish migrate over 2000 km in Yukon and Mackenzie rivers.

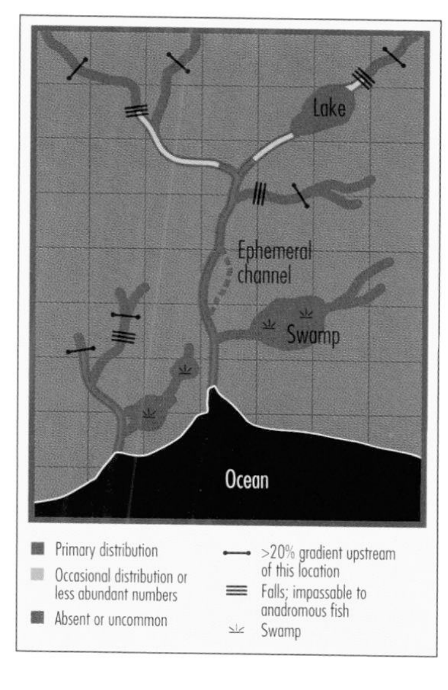

BEHAVIOUR

- Chum migrate downstream to sea soon after emergence. They are usually gone from fresh water by June 1.
- Chum may spawn in intertidal zone of small streams.

Sockeye salmon

Oncorhynchus nerka

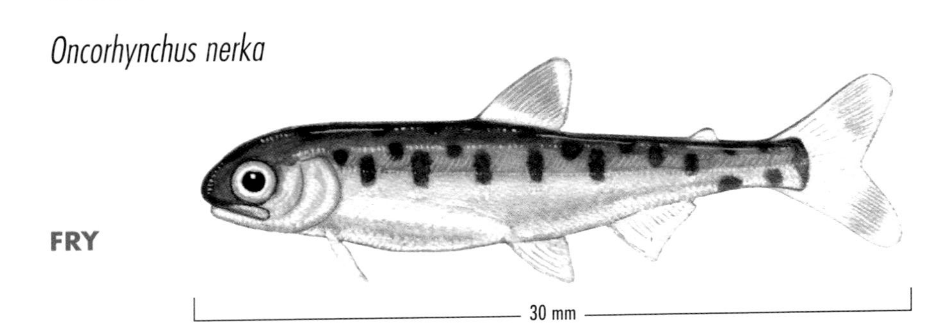

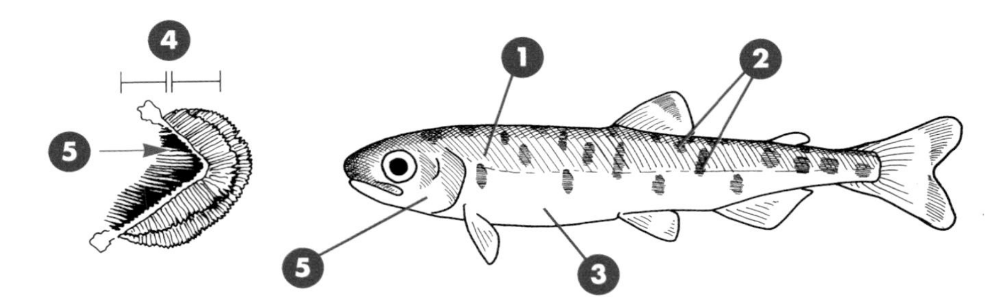

Colour & Anatomy

1. Parr mark length is less than than vertical diameter of eye.
2. Parr marks are irregular—height is irregular.
3. Area below lateral line is silver or white—no greenish sheen.
4. Gill raker length is almost = to length of gill filaments.
5. 30–39 gill rakers on first arch.

DISTRIBUTION

- Most sockeye occur in systems where young can enter a lake to rear.
- Some small populations spawn and rear in lower reaches of large rivers.
- Fry migrate to lakes from spawning areas that may be either upstream or downstream from the lake.
- Resident forms, "Kokanee," spend their entire life in fresh water.
- Sockeye rarely rear in coastal streams without a lake that is accessible to adult salmon.

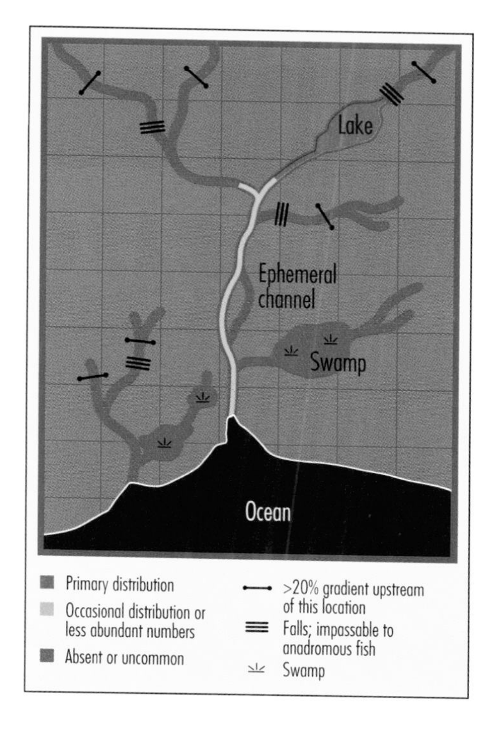

BEHAVIOUR

- In large lake-river systems, fry may migrate in great numbers along shore of the river while moving to lake.
- Sockeye may spawn on beaches in some coastal lakes.

FRESHWATER RESIDENCE TIME

- Juveniles rear in lakes for 1–2 years.

SOCKEYE VS CHUM

- Sockeye fins are larger than those of chum.
- Height of parr marks is irregular in sockeye.
- Sockeye have no greenish iridescence below lateral line.
- Chum have regular-height parr marks, usually faint or absent below lateral line, and greenish iridescence below lateral line.
- Sockeye are generally found only in watersheds with accessible lakes.
- Sockeye have long gill rakers.

Coho salmon

Oncorhynchus kisutch

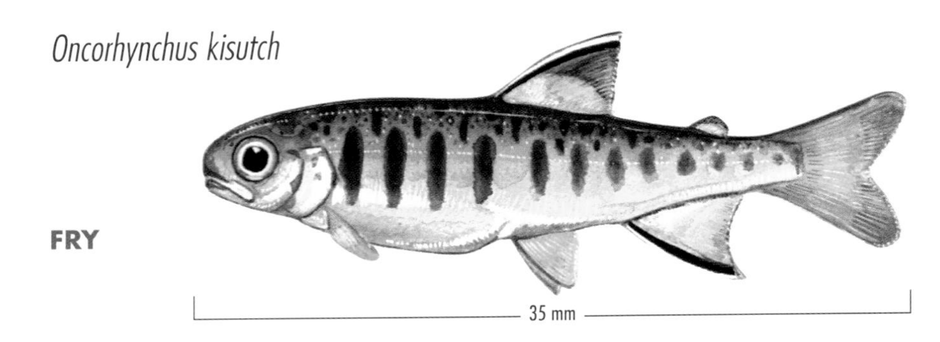

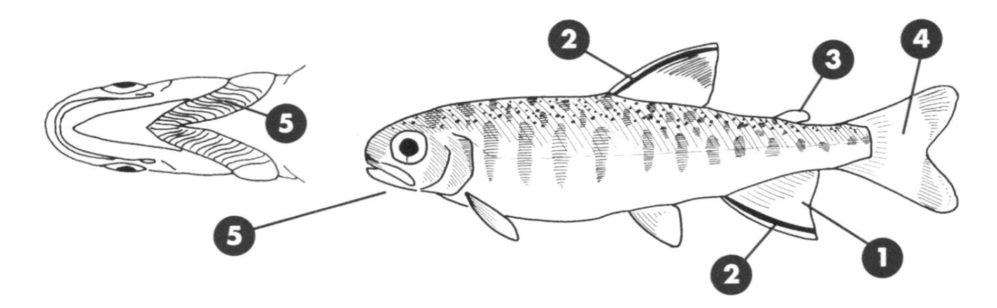

Colour & Anatomy

1 Anal fin is sickle-shaped, leading edge is longer than base.
2 Leading edges of anal and dorsal fins have white followed by black.
3 Adipose fin has dark edge, centre is opaque.
4 Caudal, anal and adipose fins are pale orange.
5 Species has 13–14 branchiostegals.
6 Species usually has 45–80 pyloric caeca.

Distribution

- Coho use all accessible reaches of streams.
- Seasonally wetted areas and off-channel sloughs, swamps and their tributaries are used for winter rearing.
- In main stems of large rivers, coho use margins, debris piles and under-cut banks.
- Coho may be found in fresh/saltwater pools in small estuaries.
- In small streams they prefer pools and glides.

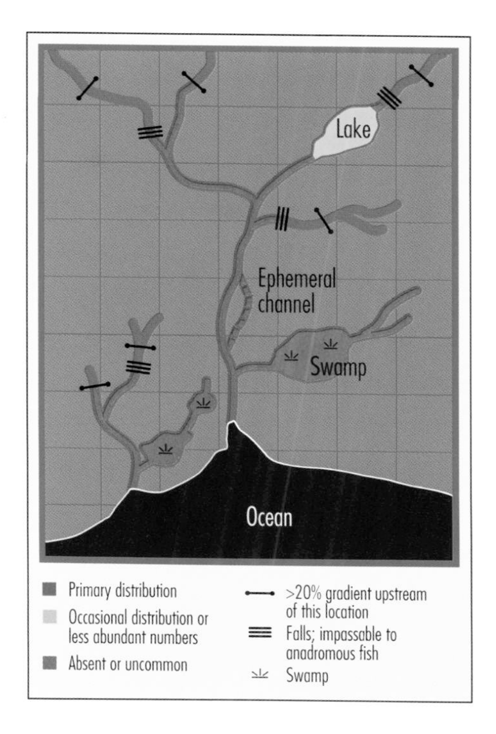

Behaviour

- Coho feed actively on anything falling into water. They are common along stream margins. Coho are the most common and easily seen salmon fry during the active growing season.
- Below about 6°C, the fry will seek cover under banks or in debris piles.
- Coho will nip and chase each other.

Freshwater residence time

- Coho remain in fresh water 1–2 years.

Other

- Coho are very widely distributed in streams of all sizes, including very small first- and second-order streams.
- Some stocks in small first- and second-order streams may not be in DFO files.
- You may encounter "Lake Forms" of coho with a less sickle-shaped anal fin.
- As coho become ready to smolt, they become brighter and the anal fin is less sickle-shaped.

Chinook salmon

Oncorhynchus tshawytscha

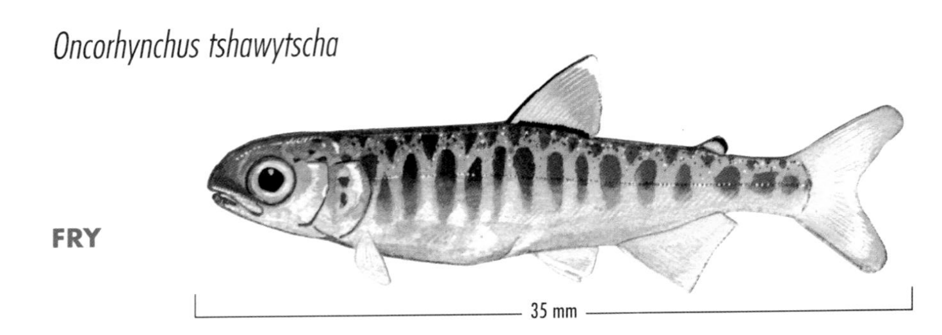

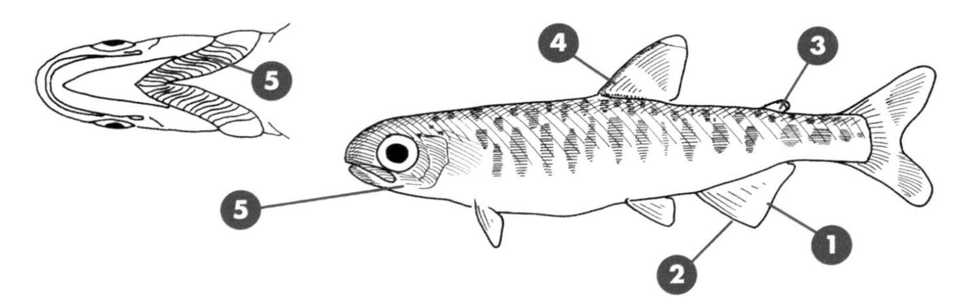

Colour & Anatomy

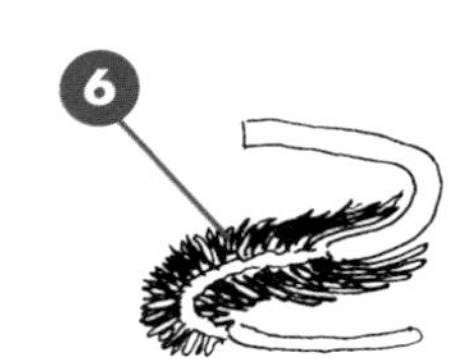

1. Anal fin is not sickle-shaped: leading edge of anal fin is shorter than length of base.
2. Anal fin leading edge is white.
3. Adipose fin has clear centre or "window."
4. Dorsal fin has dark leading edge and white tip.
5. Species has 16–18 branchiostegols.
6. Species usually has 135–185 pyloric caeca.

Distribution

- Chinook are usually found in moderate to large streams.
- Main channel is used for rearing.
- In large streams, 8–10 cm fish live in faster, deeper water than coho.
- Chinook may rear in estuaries of larger rivers—e.g. Nanaimo, Cowichan, Fraser.
- Adults may hold in lakes before spawning.

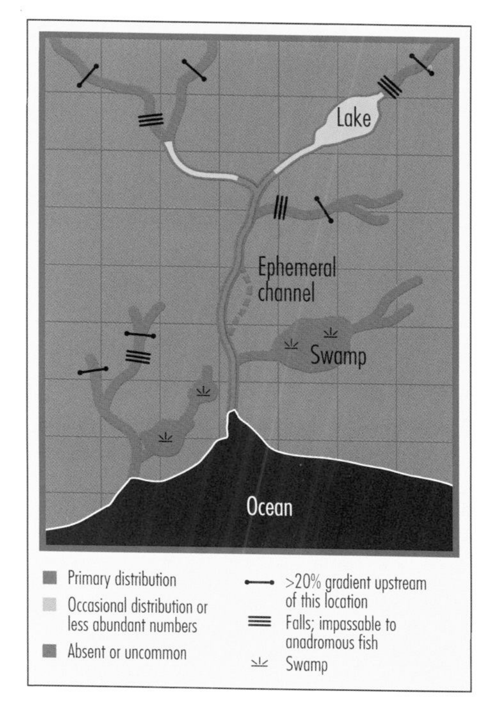

Freshwater residence time

- Chinook may form races with some rearing >1 year, some 90 days or less depending upon type. Southern populations frequently stay 90 days and northern populations are more likely to stay up to a year, but can be variable in all areas.

Other

- Most populations of chinook are known and listed in DFO escapement catalogs.
- Dorsal fin tip darkens as fish become ready to go to sea.

Chinook vs Coho

- Chinook have “clear window” in adipose.
- Chinook do not have sickle-shaped anal fins or white and black stripes on leading edges of anal and dorsal fins.
- Coho have sickle-shaped anal fins, with leading edges longer than length of base of anal fins. Leading edges of anal and dorsal fins have white and black stripes.
- Chinook frequent main stems of moderate to large rivers. Coho are found in all accessible stream reaches, including seasonally wetted areas.

Live Specimens: Dolly Varden & trout fry

Important diagnostic features visible on the photographs are identified. Refer to Identification Keys and species pages to confirm identifications made from photos. Note length of base of anal fin is less than or equal to length of base of dorsal fin on trout and char.

Length: 30 mm.

Dolly Varden Fry

Parr marks are uneven.

Length: 30 mm.

Cutthroat Fry

Cutthroat and rainbow fry are very difficult to distinguish in the field. Specimens over 80 mm *(see parr)* are usually necessary to distinguish cuthroat from rainbow.

Length: 30 mm.

Steelhead/Rainbow Fry

See cutthroat fry.

Length: 30 mm.

Brown Trout Fry

Adipose fin is orange-brown colour.

Length: 40 mm.

Atlantic Salmon Fry

Pectoral fins are long and reach to or past the insertion of the dorsal fin.

LIVE SPECIMENS: DOLLY VARDEN & TROUT PARR

Important diagnostic features visible on the photographs are identified. Refer to the Identification Charts and species pages to confirm identifications made from photos.

Length: 95 mm.

Dolly Varden Parr

Parr marks are irregularly shaped.

Fins and body have no black spots.

Length: 120 mm.

Cutthroat Parr

Length of maxillary extends past back of eye.

Adipose is spotted.

Underside of jaw has red slash mark.

White tip on dorsal fin covers 3 or fewer interspaces between rays.

Length: 100 mm.

Steelhead/Rainbow Parr

No red slash.

Maxillary does not reach past back of eye.

Adipose has black margin.

White tip on dorsal fin covers 3–5 interspaces between rays.

Length: 80 mm.

Brown Trout Parr

Adiopose has orange-brown pigment.

Red spots are present.

Pectoral fin does not reach insertion of dorsal fin.

LIVE SPECIMENS: SALMON FRY

Important diagnostic features visible on photographs are identified. Refer to Identification Charts and species pages to confirm identifications made from photos. Note: length of base of anal fin is larger than length of base of dorsal fin on all juvenile salmon.

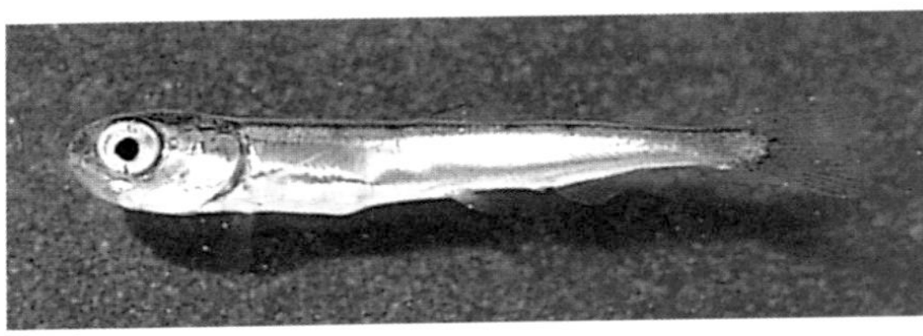

Length: 30 mm.

Pink Salmon Fry

Parr marks are absent.

Colour is very silver.

Length: 55 mm.

Chum Salmon Fry

Parr marks are even in length—more above than below lateral line.

Blue/green sheen shows below lateral line.

Length: 50 mm.

Sockeye Salmon Fry

Parr marks are uneven length, and some are equal above and below lateral line.

No green sheen shows below lateral line.

Length: 50 mm.

Coho Salmon Fry

Dorsal and anal fin margins are sickle-shaped and have white and black stripes.

Parr marks larger than eye diameter.

Fins are red or orange.

Length: 50 mm.

Chinook Salmon Fry

Dorsal has white tip.

Adipose has clear window.

Parr marks larger than eye diameter.

References

Carl, G.C., W.A. Clemens and **C.C. Lindsey**. 1959. *The freshwater fishes of British Columbia.* British Columbia Provincial Museum, Victoria, Department of Education Handbook No. 5. 192 pp.

Ellis, D.V. 1977. *Pacific salmon management for people.* Western Geographical Series, Volume 13, Department of Geography, University of Victoria, Victoria. 320 pp.

Groot, C. and **L. Margolis**. 1991. *Pacific salmon life histories.* UBC Press, Vancouver. 564 pp.

Hartman, G.F. 1956. *A taxonomic study of cutthroat trout,* Salmo clarki clarki *Richardson, rainbow trout* Salmo gairdneri *Richardson and their reciprocal hybrids.* M.A. Thesis, Department of Zoology, University of British Columbia, Vancouver. 71 pp.

McConnell, R.J. and **G.R. Snyder**. 1972. *Key to field identification of anadromous juvenile salmonids in the Pacific Northwest.* NOAA Technical Report NMFS Circ-336. U.S. Government Printing Office. Washington, D.C. 6 pp.

McPhail, J.D. and **R. Carveth**. 1994. *Field key to the freshwater fishes of British Columbia* (Draft for 1994 field testing). Prepared for Aquatic Inventory Task Force of the Resources Inventory Committee, Victoria. 233 pp.

McPhail, J.D. and **C.C. Lindsey**. 1970. *Freshwater fishes of northwestern Canada and Alaska.* Fisheries Research Board of Canada, Ottawa. Bulletin 173. 381 pp.

Phillips, A.C. 1977. *Field key characters of use in identifying young marine Pacific salmon.* Field and Marine Services Technical Report No. 746. Pacific Biological Station, Nanaimo, B.C. 13 pp.

Scott, W.B. and **E.J. Crossman**. 1964. *Fishes occurring in the fresh waters of insular Newfoundland.* Contribution No. 58 of Life Sciences, Royal Ontario Museum, University of Toronto, Toronto.

Scott, W.B. and **E.J. Crossman**. 1973. *Freshwater fishes of Canada.* Fisheries Research Board of Canada, Ottawa. Bulletin 184. 966 pp.

Trautman, M.B. 1973. *A guide to the collection and identification of presmolt Pacific salmon in Alaska with an illustrated key.* NOAA Technical Memorandum NMFS ABFL-2. U.S. Government Printing Office, Washington, D.C. 20 pp.

NOTES